AF687605

Genereris

PUBLISHING

SCIENTOMARKET:
The Architecture of Value, Trust, and Influence in Science

Bisenbaev A.K.

Title: **SCIENTOMARKET:**
The Architecture of Value, Trust, and Influence in Science

ISBN: 979-8-89966-273-7

Author: Bisenbaev A.K.

Cover image: AI Generated

Publisher: Generis Publishing
Online orders: www.generis-publishing.com
Contact email: info@generis-publishing.com

Bisenbaev A.K. *Scientomarket: The Architecture of Value, Trust, and Influence in Science. GENESIS*, 2025.

The monograph *Scientomarket* presents an original philosophical and methodological study of a new disciplinary framework—scientific marketing (Scientomarket)—which approaches science not as the production of knowledge but as an architecture of value, trust, and influence.

We argue that the crisis of today's scientific world lies not in a lack of publications but in the devaluation of trust. Texts proliferate faster than their verifiability does, and social significance can consolidate.

For the first time, this work proposes a systemic theory of the scientific market of meanings—a space where not capital but truth circulates and where not advertisement but evidential rigor defines success. In the notation of *Scientomarket*, we describe this as the engineering of the conversion cycle "value–trust–influence–resources"—a managed circuit of reproducible scientific growth.

Thus, our framework unites metascience, scientometrics, institutional economics, and epistemology, translating the language of science into the economy of flows and trust infrastructures.

A scientomarket, in this sense, is an invitation to return to science its genuine market—the market of truth and reproducible evidence.

This work focuses on philosophers of science, metascience researchers, policymakers in R&D, heads of research institutions, and everyone who seeks to comprehend science as a self-developing ecosystem of trust and influence.

This research is funded by the Science Committee of the Ministry of Science and Higher Education of the Republic of Kazakhstan (Grant No. AP26199077).

General Acronyms and Frameworks

Abbreviation	Full Term	Meaning/Function
OA	Open Access	A publishing model providing free access to research outputs.
PRISMA	Preferred Reporting Items for Systematic Reviews and Meta-Analyses	Standard for transparent and complete reporting of systematic reviews.
CONSORT	Consolidated Standards of Reporting Trials	Reporting standard for randomized controlled trials.
CTO	Chief Technology Officer	Industrial stakeholder overseeing technological translation.
FAIR	Findable, Accessible, Interoperable, Reusable	Principles for scientific data management and stewardship.
DORA	San Francisco Declaration on Research Assessment	Global initiative to improve the evaluation of scientific output.
REE	Reputation Elasticity to Evidence	Rate at which trust increases with accumulation of verifiable evidence.
IHL	Influence Half-Life	Time during which the influence of a result decays by half if not renewed by further evidence.
KCR	Knowledge Conversion Ratio	Proportion of results that successfully move from one stage of the scientific circuit to the next.
K-ROI	Knowledge Return on Investment	Aggregate efficiency of creating and distributing scientific products.
SPMF	Science Product–Market Fit	Degree of correspondence between the evidential strength, utility, cost, time,

		and risk of a knowledge product for a given segment.
EME	Ethical Marketing Envelope	Integrity framework defining ethical and communicative "red lines" in scientomarketing.
MOOC	Massive Open Online Course	Digital educational platform for disseminating scientific knowledge.
MVKP	Minimal Viable Knowledge Product	The minimal reproducible and verifiable form of a scientific result (text + data + code + protocol).
ESpec	Evidence Specification	Structured documentation of data, methods, and artifacts ensuring reproducibility.
KBOM	Knowledge Bill of Materials	Comprehensive list of components (datasets, code, protocols) forming a reproducible research package.
SRM	Science Relationship Management	Analogous to CRM; system for managing interactions and communication in scientific ecosystems.
TTI	Time-to-Impact	Duration between publication and measurable influence or adoption of a result.
BMJ	British Medical Journal	Example source cited for PRISMA/CONSORT.
HEFCE	Higher Education Funding Council for England	Referenced in *The Metric Tide* report.

Table of contents

Introduction: Why Science Needs Its Own Market

It is becoming increasingly evident that the modern scientific system has entered an era of paradoxes. Never before has the production of scientific results been so large-scale and rapid - and yet never before has their social validity and long-term value been so questioned.

Science today exists amid simultaneous abundance and scarcity: texts multiply faster than trust accumulates; citations rise faster than truth is confirmed; new journals appear faster than standards of quality stabilize. This gap between the speed of circulation and the speed of verification is, we believe, the defining symptom of the current crisis in the scientific ecosystem.

Hence, it becomes clear that the habitual optics of *"science as knowledge production"* are insufficient. Knowledge production is merely one layer of a larger structure. Science is something greater: it is an architecture of value, trust, and influence. It operates not only through the generation of facts and theories but also through the complex circuits of their testing, recognition, dissemination, reputation building, and long trajectories of impact on other domains of knowledge and social practice.

If economies require markets for goods and capital, science requires its own market of meanings - a structured space of epistemic exchange where the circulating unit is not monetary equivalence but verifiable truth, methodological rigor, evidential beauty, and institutional reliability.

In this work, we articulate this expanded space of science through a new term - Scientomarket.

Today's science faces a structural erosion of trust. On the one hand, the "replication crisis" across several disciplines has exposed the fragility of many findings and practices[1]. On the other hand, an industrially accelerated publication market has spawned weak peer reviews, "turbo-editing," and surrogate signals of quality[2].

---

[1] Nosek, B. (2015). Open Science Collaboration: Estimating the reproducibility of psychological science. Science, 349(6251), aac4716–aac4716.

[2] Shen, C., & Björk, B. C. (2015). "Predatory" open access: A longitudinal study of article volumes and market characteristics. BMC Medicine, 13(1). Doi: 10.1186/s12916-015-0469-2

The publication overload is accompanied by attention inflation: reviewers and readers are time constrained, journals compete for "novelty," and institutions compete for quantifiable outputs. In this context, citation - originally a marker of influence - has become an ambiguous indicator, often reflecting successful embedding in communication flows (thematic alignment, network effects, headline design, preprint accessibility) rather than the robustness of results[3].

Consequently, the critique of "metrics without context" grows loudly - from the *Leiden Manifesto* to *The Metric Tide*[4] and from Seglen's[5] classic argument against the impact factor as a universal measure of quality.

We emphasize that the problem does not lie in *open access* per se. Indeed, open access accelerates exchange and lowers barriers[6]. The core risk lies instead in the mismatch between institutional capacity and informational velocity. When the emission of papers outpaces the formation of verification infrastructures, trust is inevitably debased. When "indices" and "rankings" substitute for substantive evidence, a citation ceases to be a synonym for truth[7].

Therefore, modern science needs a different descriptive language - a *sci-economic* one that captures not only products (papers) but also conversion processes: how epistemic value becomes trust and how trust becomes influence.

We interpret this new language through three interconnected theses:

1. Value in science is neither market price nor social fame but rather the capacity of a result to reduce uncertainty and expand the explanatory space. Value resides in protocols, data, code, and methods - that is, in everything that makes results transferable and verifiable[8].
2. Trust is not a mood or the charisma of the author but an institutionally secured confidence - a constellation of procedures (peer review, artifact openness,

---

[3] Bornmann, L., & Daniel, H. D. (2008). What do citation counts measure? A review of studies on citing behavior. Journal of Documentation, 64(1), 45–80. https://doi.org/10.1108/00220410810844150
[4] Hicks, D., Wouters, P., Waltman, L., De Rijcke, S., & Rafols, I. (2015, April 23). Bibliometrics: The Leiden Manifesto for research metrics. Nature. Nature Publishing Group. https://doi.org/10.1038/520429a
[5] Seglen, P. O. (1997). Why the impact factor of journals should not be used for evaluating research. British Medical Journal, 314(7079), 498–502. https://doi.org/10.1136/bmj.314.7079.497
[6] Piwowar, H., Priem, J., Larivière, V., Alperin, J. P., Matthias, L., Norlander, B., … Haustein, S. (2018). The state of OA: A large-scale analysis of the prevalence and impact of Open Access articles. PeerJ, 2018(2). https://doi.org/10.7717/peerj.4375
[7] Moed, H. F. (2009, February). New developments in the use of citation analysis in research evaluation. Archivum Immunologiae et Therapiae Experimentalis. https://doi.org/10.1007/s00005-009-0001-5
[8] Ioannidis, J. P. A. (2018). Why most published research findings are false. In Getting to Good: Research Integrity in the Biomedical Sciences (pp. 2–8). Springer International Publishing. https://doi.org/10.1371/journal.pmed.0020124

replication, ethical standards) that transforms private claims into public knowledge[9].
3. Influence is not reducible to citation counts. It is the reconfiguration of practices - the inclusion of methods in curricula, the emergence of standards, the diffusion of code libraries and benchmark datasets, and the redirection of research agendas[10].

In short, science functions as a system of meaning exchange, where value–trust–influence forms its load-bearing triad. The value without trust remains a local contribution; trust without value is an empty sign; influence without both is media noise. Only their coherence constitutes the "capital" of the scientific world - durable results that others can understand, verify, and apply[11].

The metaphor of "science as an economy" is not about monetizing knowledge but rather about translating it into the language of flows, institutions, and costs[12] . In this ecosystem, meanings and verifiable truths act as capital; evidential claims and reproducible methods act as goods; and journals, repositories, indices, peer-review mechanisms, and educational and regulatory bodies act as the financial infrastructure[13].

As in economic markets, there are transaction costs (to understand, verify, reproduce), information asymmetries (access to data and code), clearing mechanisms (editorial and review processes), and risks of inflation (accelerated paper emission without equivalent verification). Thus, a practical imperative arises: to increase the liquidity of trust so that sound knowledge finds its audience quickly, while dubious work does not monopolize attention[14].

The very idea of *Scientomarket* emerged at the intersection of three trajectories:

First, the critique and renewal of scientometrics. Classical bibliometric indicators endow science with unprecedented mapping tools yet reveal their limits

---

[9] Resnik, D. B., & Master, Z. (2013). Policies and Initiatives Aimed at Addressing Research Misconduct in High-Income Countries. PLoS Medicine, 10(3). https://doi.org/10.1371/journal.pmed.1001406

[10] Fortunato, S., Bergstrom, C. T., Börner, K., Evans, J. A., Helbing, D., Milojević, S., ... Barabási, A. L. (2018, March 2). Science of science. Science. American Association for the Advancement of Science. https://doi.org/10.1126/science.aao0185

[11] Daston, L., & Shapin, S. (1995). A Social History of Truth: Civility and Science in Seventeenth-Century England. The Journal of Philosophy, 92(7), 388. https://doi.org/10.2307/2941063

[12] Porter, T. M. (2020). The Political Philosophy of Quantification. In Trust in Numbers (pp. 73–86). Princeton University Press. https://doi.org/10.23943/princeton/9780691208411.003.0005

[13] Larivière, V., Haustein, S., & Mongeon, P. (2015). The oligopoly of academic publishers in the digital era. PLoS ONE, 10(6). https://doi.org/10.1371/journal.pone.0127502

[14] Wilsdon, J. (2017). The Metric Tide: Independent Review of the Role of Metrics in Research Assessment and Management. The Metric Tide: Independent Review of the Role of Metrics in Research Assessment and Management. SAGE Publications Ltd. https://doi.org/10.4135/9781473978782

when they are turned from means into ends. The new wave of research stresses contextual citation profiles and multidimensional influence patterns.

Second, institutional evolution. The rise of open science - preprints, repositories, open data and codes, registered reports, and postpublication reviews - has generated a multilayered infrastructure of exchange[15], deepening the dependence on stable platforms and distributed quality-control practices.

The third category includes socioepistemological studies of science. From laboratory ethnography to the "social history of truth," it became clear that truth in science is not only a logical but also an organizational category, relying on trust practices, networks, and material carriers.

The synthesis of these trajectories yields *Scientomarket*. Here, we shift the focus from counting traces to analyzing architectures of exchange - from articles as end products to conversion chains (value - trust - impact) sustained by specific institutions and standards.

We move from scientometrics as accounting to sci-economics as flow management, where the principal objects are nodes and channels of trust, verification costs, and the transport of knowledge across disciplines.

What, then, does the *Scientomarket* perspective achieve?

First, it shifts the discussion from static indicators to trajectories. What matters is not how often a text is cited "now" but how it is verified, which artifacts are open, how swiftly and where independent applications emerge.

Second, it demands infrastructures of transparent accountability - open peer reviews, standardized protocols, open data and code, stable replication practices, and institutional recognition[16].

Third, it redefines the role of institutions (editors, funders, universities): they become operators of meaning markets, maintaining the liquidity of trust, reducing verification costs, and preventing the inflation of surrogate signals.

Finally, the *Scientomarket* perspective restores rhetoric and designs its status as internal components of evidence. *How* something is said affects *how* it is understood

---

[15] Nosek, B. A., & Lakens, D. (2014). Registered reports: A method to increase the credibility of published results [Editorial]. Social Psychology, 45(3), 137–141. https://doi.org/10.1027/1864-9335/a000192

[16] Resnik, D. B., & Master, Z. (2013). Policies and Initiatives Aimed at Addressing Research Misconduct in High-Income Countries. PLoS Medicine, 10(3). https://doi.org/10.1371/journal.pmed.1001406

and *how* it is verified. A well-designed scientific text is thus not cosmetic - it is an interface of trust (Tufte, 2006; Hicks et al., 2015).

In conclusion, *Scientomarket* is an invitation to return science to its own market - a market not of money but of truth; not of flashy announcements but of reproducible proofs; not of monologs but of institutionally guaranteed dialogs.

If the scientific system is to emerge from its paradoxical age, it needs not "more metrics" but better rules of exchange - visible chains of conversion from value to trust and influence, visible statuses of verifiability, and infrastructures that make trust liquid and influence structure.

Chapter 1. What Is Scientomarket

We begin with a few self-evident truths.

Scientific results are born in local laboratories, universities, institutes, and theoretical schools—or, at times, elsewhere: in a home office, or even in a dream. However, to become part of collective knowledge, these results must travel a long and uncertain path: from an initial claim to an accepted fact, from local verification to external replication, and from publication to application and standardization.

Along this path unfolds a chain of crucial transformations - namely, the conversion of epistemic value into trust, of trust into influence, and of influence into resources. These resources are then reinvested to increase the quality and openness of future research.

This relatively simple causal chain, in fact, constitutes the circuit of reproducible scientific growth. Within this context, our new concept – Scientomarket - is defined as the discipline of *designing and managing* that very circuit.

We do not mean by this the "PR of science," science popularization, or the selling of scientific products. Rather, *Scientomarket* is understood as the engineering of correspondence between what has been proven, who needs it, and how it should be packaged and transmitted through the institutions of trust to become effective public action.

1.1 Definition, Subject, and Mission

Accordingly, we define *Scientomarket* as a system of strategic and operational practices through which scientific teams, institutions, and research ecosystems manage the creation, validation, translation, and distribution of scientific value among diverse stakeholders - editors and reviewers, funding bodies and ministries, industry and professional associations, educational platforms, and the public at large.

From our own scientific practice, we have learned that such issues are crucial to the entire process of scientific production, yet they remain poorly theorized and weakly institutionalized.

In some regional scientific communities, one often hears the following question: *"Do we even have science?"* Of course, some form of science always exists. Such questions usually arise not from their absence but from the total lack of communication and interpretation between society and science. They have become so disconnected that they barely perceive one another.

Applied science - or its commercialization - often becomes stuck somewhere along the way: between idea and competence, interpretation and editorial review, peer review and funding, grant programs and ministries, industries and associations, local authorities and communication platforms. The result rarely reaches the broader public.

Therefore, we view *Scientomarket* precisely as this system of strategic and operational practices that can enable or assist national scientific systems in translating their ideas, projects, and products into a *state of percolation* - that is, continuous diffusion - on their way toward society and the economy. Naturally, we cannot cover every issue in detail, but we shall at least outline the key contours of such percolation and the design of a *"green corridor"* between science and society.

Overall, our framework rests on three pillars:

1. Evidential rigor[17],
2. Institutions of reproducibility and openness[18],
3. The trust-based economy of "credence goods"[19].

Within this notation, the mission of *Scientomarket* is not to "sell" science to society but to reduce the friction between truth and action - to accelerate the path from discovery to effect, without sacrificing epistemic rigor.

Second, it seeks to align results with the needs of their respective segments (journals, regulators, industries) without replacing causality with convenient storytelling.

Third, it aims to transform fragmented communications into managed transition funnels, establishing measurable indicators and ethical constraints at each stage.

---

[17] Namer, G., Merton, R. K., Storer, N. W., & Namer, G. (1975). The Sociology of Science. Theoretical and Empirical Investigations. Revue Française de Sociologie, 16(1), 144. https://doi.org/10.2307/3321144

[18] Munafò, M. R., Nosek, B. A., Bishop, D. V. M., Button, K. S., Chambers, C. D., Percie Du Sert, N., … Ioannidis, J. P. A. (2017, January 10). A manifesto for reproducible science. Nature Human Behaviour. Nature Publishing Group. https://doi.org/10.1038/s41562-016-0021

[19] Gambetta, D. (2000). Can We Trust Trust? Trust: Making and Breaking Cooperative Relations. Department of Sociology, University of Oxford, (chapter 13), 213–237.

1.2 Why It Is Not PR or Popularization

We have ourselves been involved in both corporate and personal PR, as well as in the popularization of science within a national context. To clarify, PR constructs recognition and sustains the reputational background, whereas popularization educates and engages broad audiences. Both domains are, of course, important - but they operate *around* or *after* the evidential core.

Scientific marketing, however, follows a fundamentally different logic. It begins *inside* the scientific product - within protocols, data, code, and reproducible artifacts - and then builds bridges from this evidential nucleus to various regimes of acceptance.

Specifically, a PR can generate *visibility without the transfer of agency*, whereas popularization can create *interest without interfaces for application*. In contrast, Scientomarket insists on a strict principle: no promises beyond the data, and no accelerators that bypass verification institutions.

In this sense, *Scientomarket* is designed to increase speed not through rhetoric but through the engineering of verifiability, transferability, and compatibility.

After all, in organizational-theory terms, science addresses a very particular kind of exchange object - not a commodity in the usual sense but a claim with evidence - the quality of which the end-user often cannot fully evaluate. These objects are known as credence goods.

Consequently, institutions of trust - peer review, open data and code, replication studies, and standards of reporting and exchange - are not external "administration" but rather the arteries through which scientific value circulates[20].

Hence, we see the task of *Scientomarket* precisely as the design of this traffic of value.

1.3 The cycle of value–trust–influence–resources

For greater clarity, let us formalize the process in a simple procedural model.

[20] Page, M. J., McKenzie, J. E., Bossuyt, P. M., Boutron, I., Hoffmann, T. C., Mulrow, C. D., … Moher, D. (2021, March 29). The PRISMA 2020 statement: An updated guideline for reporting systematic reviews. BMJ. BMJ Publishing Group. https://doi.org/10.1136/bmj.n71

At the input, we have *epistemic value* - an explanation, a prediction, a new method, or a new dataset. However, value does not exist alone because it becomes activated only when it passes through mechanisms that generate trust: open data and code, preregistration, independent replication, and rigorous reporting (for example, within the frameworks of FAIR, CONSORT, or PRISMA).

Trust, in turn, is converted into influence - manifested as citations in reviews and standards, inclusion in educational programs, or use in technological solutions and policies.

Influence then generates resources - grants and contracts, access to infrastructure and data, and the inflow of talent. Finally, resources return to research as amplifiers of quality and openness, thus creating a positive feedback loop.

We can thus interpret this logic as a cascade with a reinvestment loop - a dynamic circuit in which the production, verification, and dissemination of knowledge form a self-sustaining system of epistemic growth.

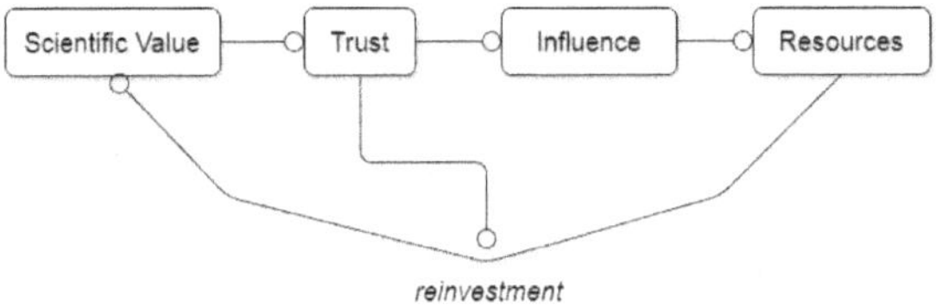

In a minimal formalization, such a circuit can be described by a state-transition function whose passages depend on the quality of evidence, the costs of implementation, and temporal lags.

We need this scheme not for formalism's sake but for governability. In this way, *Scientomarket* translates each arrow in the diagram into a work plan and metrics.

Let us now introduce the minimal indicators to which we will continually return:

Influence Half-Life (IHL) - the time over which the influence of a result halves if it is not "replenished" with new evidence and implementations,

Reputation Elasticity to Evidence (REE) - the rate at which trust grows as open and verifiable evidence accumulates,

Knowledge conversion ratio (KCR) - the share of results that transition from one stage of the circuit to the next (for example, "*disclosure - external verification,*" "*acceptance - implementation*"),

Knowledge return on investment (K-ROI) - the integral profitability of assembling and distributing the scientific product.

By these indicators, we mean not a "KPI dashboard" but a map of bottlenecks. For example, a low KCR almost always signals poor specification of artifacts (data, code, protocols). A high decay of influence (small IHL) typically indicates a lack of replications and transferable case studies. A low K-ROI is interpreted as "visibility without application" (which may be called "parade science"). By *parade science,* we refer to national research systems that officially proclaim scientific success even as, by most or nearly all indicators, science is objectively degrading—in short, when official subjectivity refuses to reconcile with factual objectivity.

1.4. Stakeholder Map of Science: From Actors to Regimes of Acceptance

Because science does not address a "*market in general*" but rather specific audiences, it is important to distinguish regimes of acceptance. We divide them into:

Academic (editors, reviewers, readers),

Policy-regulatory (ministries, agencies, standards committees),

Industrial (R&D units, CTOs, product teams),

Educational (universities, MOOC platforms), and

Public media (mass media, science journalism, civic communities).

Each regime, in our notation, demands its own "ladder of evidence" and relies on distinct channels - for example, journals and preprints, data and code repositories, policy briefs and standards, pilots and prototypes, curricula and teaching sets, visualizations and AMA sessions.

It is convenient to hold this plurality in a single picture—namely, as a stakeholder map

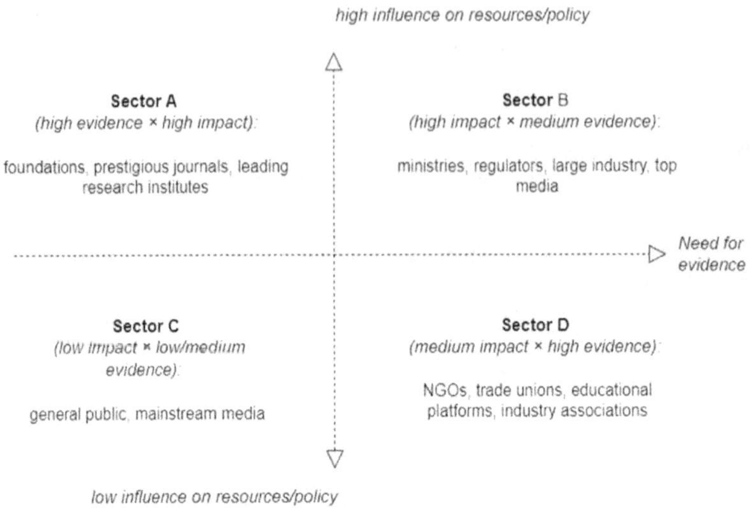

On this map, along the horizontal axis, we plot the required evidential level (from popular formats to systematic reviews and replications). Along the vertical axis, we plot the influence on resource allocation (from soft legitimation to decisions on funding priorities and regulatory change).

In the upper-right quadrant, we place funding bodies and prestigious journals, regulators and standard-setting organizations. In the lower-left quadrant, we place mass media and a broad public audience. Among them lie industry associations, educational platforms, and professional communities.

The transitions between these sectors are precisely the strategic funnels of *Scientomarket*: from academic verification to a policy package; from a pilot to a standard; and from an open course to practical embedding in professional routines.

1.5. The Boundaries of the Discipline and Its "Red Lines"

Importantly, our concept of *Scientomarket* neither replaces scientific methodology nor promises to accelerate truth through rhetoric, nor does it engage in lobbying. Its "red lines" lie primarily within the ethical envelope of communication.

By this, we mean a clear labeling of a result's status (hypothesis, preliminary effect, replicated effect, standard); transparent disclosure of conflicts of interest and sources of funding; and a categorical refusal of "stretched causality" where only correlations are available[21].

We refer to this integrity framework as the ethical marketing envelope (EME).

Without EME, any increase in speed becomes a risk of rupturing institutional bonds of trust.

This problem is particularly acute in certain conservative or incompetent scientific systems that dislike long-term projects. They want *everything at once and as fast as possible*. Such impatience produces precisely what we previously called cargo science or "parade science" - a simulacrum of research that imitates productivity while undermining epistemic reliability.

Conversely, *Scientomarket* deliberately assumes responsibility for everything that lies between primary truth and implementation:

- designing the *minimally viable knowledge product* (article, dataset, code, protocol, license),
- Ensuring containerization and reproducibility,
- interfacing with existing standards,
- preparing *policy-ready* and *industry-ready* packages,
- designing pilots and traceable effect reporting, and
- establishing a clear economy of metrics (TTI, IHL, KCR, K-ROI) without which strategic promises remain words and declarations.

1.6. Lessons from Metascience: How Institutions Shape Trajectories

Histories of reproducibility crises across disciplines[22] have revealed a simple truth: institutions matter.

---

[21] Strathern, M. (1997). 'Improving ratings': audit in the British University system. European Review, 5(03), 305. https://doi.org/10.1017/s1062798700002660

[22] Klein, R. A., Vianello, M., Hasselman, F., Adams, B. G., Adams, R. B., Alper, S., ... Nosek, B. A. (2018). Many labs 2: Investigating variation in replicability across samples and settings. Advances in Methods and Practices in Psychological Science, 1(4), 443–490. https://doi.org/10.1177/2515245918810225

The FAIR principles made data findable and reusable[23]. CONSORT and PRISMA established reporting standards for clinical trials and systematic reviews[24]. Open Science normalized preregistration and code publication.

These practices collectively increase the Reputation Elasticity to Evidence (REE) - the responsiveness of trust to new proof - and slow the Influence Half-Life (IHL) - the decay rate of impact over time. Each additional *quantum of evidence* thus becomes more visible, more easily verifiable, and hence more convertible into acceptance.

Here, the question of fit between the scientific product and the "knowledge market" arises.

At the conceptual level, this can be expressed by a simple function of science product–market fit (SPMF) that links evidential strength, usefulness for a specific segment, time to effect, implementation costs, and risk.

Let

E - integral measure of evidence (design quality, data/code openness, replications, external verifications),

U - expected utility or problem-solving capacity for the target segment,

T - time horizon to effect,

C - aggregate costs of adaptation and implementation,

R - risk (methodological, reputational, regulatory).

Then, the correspondence index can be defined as:

$$SPMF = \alpha E + \beta U - \gamma C - \delta R - \kappa T \quad (1)$$

where the coefficients reflect the weights specific to each segment.

[23] Wilkinson, M. D., Dumontier, M., Aalbersberg, Ij. J., Appleton, G., Axton, M., Baak, A., … Mons, B. (2016). Comment: The FAIR Guiding Principles for scientific data management and stewardship. Scientific Data, 3. https://doi.org/10.1038/sdata.2016.18

[24] Schulz, K. F., Altman, D. C., & Moher, D. (2010). CONSORT 2010 Statement: Updated guidelines for reporting parallel group randomised trials. Italian Journal of Public Health, 7(3), 325–332. https://doi.org/10.4178/epih/e2014029

The significance of this formula lies not in illusory precision but in the discipline of thinking.

Any plan to "bring a knowledge product to audience X" must honestly answer what we are doing to increase *E* and *U* and what we are doing to reduce *C*, *R*, and *T*.

Improving only one parameter may fail. Industry, for example, is as sensitive to *T* and *C* as it is to *E*.

Editorial boards of elite journals, by contrast, often increase acceptance probability almost linearly with *E*, while reacting only weakly to *U* unless it aligns with their internal agenda.

In this sense, the work of *Scientomarket* begins precisely where we stop merely telling stories and start engineering these variables—through the design of E-Spec/KBOM structures, pilots, adaptive packages, standardizable interfaces, and transparent public reporting formats.

We may observe that such an engineering attitude toward the scientific product becomes especially evident in typology.

A scientific paper *is* a product - but neither the only nor always the primary one.

Data become a product when accompanied by metadata, schemas, versioning, and licensing that ensure reliable reuse,

Software code is a product if it is buildable and testable - not merely "lying somewhere in a personal repository",

A protocol is a product if it truly allows the declared result to be reproduced on different equipment and with a different sample, not merely "under our conditions on our setup."

Theories and models form a special class of products: they generate explanatory and predictive power, but only if accompanied by open mathematics, sound assumptions, canonical examples, and – ideally - reference implementations.

A prototype and a patent demarcate research from technology, but this separation must not become a rupture. If one cannot move from paper to protocol, from protocol to code, or from code to prototype, then what stands before us is not a product chain but a gap between cognitive and operational value.

Thus, in our interpretation, *Scientomarket* serves as the language by which science can speak to the world - a language in which truth becomes action and action returns as resources to sustain the next truth.

Chapter 2. Actors and Scientific "Markets"

Scientific knowledge does not converge into a single point; rather, we may say that it diverges in fans. The same result - accompanied by its data, code, and protocols - simultaneously "knocks" multiple doors: the editorial policies of journals, the agendas of funding agencies, the technological roadmaps of industry, the discussions of regulators, and the curricula of educational institutions.

Each of these arteries has its own gates, queues, admission rules, and recognizable signals of validity. Therefore, any discussion of the actors of science cannot be conducted in the register of an abstract "public." The market of knowledge is multifaceted: it is a space of exchange among authors, reviewers, editors, funding bodies, corporations, states, standardization agencies, educational platforms, and the media.

Economists describe such a construction as a multisided market in which platforms - journals, databases, repositories, conferences, regulatory forums - bring different sides together, generating network effects and cross-subsidies[25]. However, science has its own specificity. Its "commodity" is not an object but a claim accompanied by evidence - a *credence good*[26]. Whose quality can seldom be fully evaluated by the end-user at the moment of exchange. Consequently, institutions of trust (peer review, open data and code, reproducibility, reporting and exchange standards) are not secondary oversight mechanisms - they constitute the infrastructure of the market itself.

2.1 Who "Buys" Knowledge and What Exactly They Buy

From the standpoint of the decision-making agent, knowledge is not merely a text. It is a bundle of claims and validations - theory, data, code, and protocol - encapsulated in interfaces that allow different operations:

- *to reproduce* (academia),

[25] Rochet, J. C., & Tirole, J. (2003). Platform competition in two-sided markets. Journal of the European Economic Association, 1(4), 990–1029. https://doi.org/10.1162/154247603322493212
[26] Dasgupta, P., & David, P. A. (1994). Toward a new economics of science,' Research Policy, 23, 487–521. Research Policy, 23(5), 487–521.

- *to translate into solutions* (industry, regulators),
- *to embed into norms* (standardizers), and
- *to disseminate* (education, media).

There are five major classes of demand:

1. Epistemic demand - the academic system (editors, reviewers, readers) seeks *evidential novelty*, *reproducibility*, *openness of artifacts*, and *logical precision*. Here, replications, preregistration, FAIR data, and reporting standards such as CONSORT and PRISMA are critical for corresponding designs,
2. Policy-regulatory demand - ministries, agencies, standards committees, parliaments. Here, value lies in translating evidence into normative action: systematic reviews, risk–benefit assessments, scenario analyses, and standardization packages. Traceability of causality and explicit accounting for uncertainty are essential (GRADE, PRISMA),
3. Innovation-technological demand — corporate R&D departments and startups. They seek to reduce risks, accelerate time-to-market, and differentiate products. Pilots, prototypes, interoperable data formats and code libraries, licensing, and shared intellectual property play decisive roles[27],
4. Educational demand - universities, colleges, MOOC platforms, and professional associations. They require educational datasets, laboratory modules, and courses that translate results into stable educational practices[28],
5. Public-media demand - newsrooms, science journalists, and professional communities. The goal here is understanding without distortion, honest articulation of applicability boundaries, and resistance to infodemics[29].

Each class sets a different threshold of evidence and uses different channels. For our purposes, this means that *Scientomarket* only begins to untangle the knot when these thresholds and channels are mapped explicitly - not held vaguely "*in mind*".

2.2 Market Failures in the Knowledge Market

We may confidently assert that knowledge markets contain built-in failures.

[27] Bogers, M., Chesbrough, H., & Moedas, C. (2018). Open innovation: Research, practices, and policies. California Management Review, 60(2), 5–16. https://doi.org/10.1177/0008125617745086
[28] Stephan, P. (2012). How Economics Shapes Science. How Economics Shapes Science. Harvard University Press. https://doi.org/10.4159/harvard.9780674062757
[29] Lewandowsky, S., Ecker, U. K. H., Seifert, C. M., Schwarz, N., & Cook, J. (2012). Misinformation and Its Correction: Continued Influence and Successful Debiasing. Psychological Science in the Public Interest, Supplement, 13(3), 106–131. https://doi.org/10.1177/1529100612451018

Information asymmetry produces *lemons* - overvalued findings that prove difficult to reproduce[30]. Metrics, when turned into goals, distort behavior.

Replications are underfunded because they constitute a public good[31]. Translation often stalls at the boundaries of standards and interfaces[32].

In this landscape, institutional technologies - FAIR data principles, CONSORT/PRISMA frameworks, open science practices, the DORA and the *Leiden Manifesto*, mission-oriented programs - act as repair mechanisms.

We argue that these mechanisms increase reputation elasticity to evidence (REE), reduce transition costs (thus improving KCR), and protect causality from simultaneous influence.

2.3 Diffusion and Norms: Where Good Ideas Go

Why do some results rapidly collapse their temporal lags while others stand in line for recognition for years?

The literature on innovation diffusion[33] and the economics of science offers a clear answer: the decisive factors are channels, norms, and combinatorial compatibility.

Results that embody compatibility - datasets in FAIR formats, code in sustainable ecosystems, protocols aligned with regulatory frameworks - diffuse faster, regardless of the strength of the original effect.

Sometimes this suffices to overtake the agenda from louder but poorly packaged work.

From this perspective, *Scientomarket* is not an "applied tail" of science but its physics of compatibility - the engineering of friction and flow. It measures

---

[30] Ioannidis, J. P. A. (2018). Why most published research findings are false. In Getting to Good: Research Integrity in the Biomedical Sciences (pp. 2–8). Springer International Publishing. https://doi.org/10.1371/journal.pmed.0020124

[31] Arrow, K. J. (1962). Economic welfare and the allocation of resources for invention. he rate and direction of inventive activity: Economic and social factors. Princeton University Press, 609–626. Retrieved from http://www.nber.org/books/univ62-1

[32] Cohen, W. M., Nelson, R. R., & Walsh, J. P. (2002). Links and Impacts: The Influence of Public Research on Industrial R&D. Management Science, 48(1), 1–23. http://www.jstor.org/stable/822681

[33] Rogers, E. M., Singhal, A., & Quinlan, M. M. (2019). Diffusion of innovations. In An Integrated Approach to Communication Theory and Research, Third Edition (pp. 415–433). Taylor and Francis. https://doi.org/10.4324/9780203710753-35

environmental resistance (coordination costs, institutional lags) and purposefully reduces it.

Thus, for a laboratory, a "correct map" means not a list of random contacts but a strategy of alignments: *Who are our A-segments (high evidence × high influence)? What packages do they need? Where do we lead the funnel tomorrow morning?*

For a university, communications are assembled into an operational model: an SRM (science relationship management) system, regular publication rhythms, A/B testing of formats, feedback journals, and departmental dashboards tracking KCR, TTI, IHL, and K-ROI.

For a funding body, programs are designed as mission-oriented funnels, with predefined interfaces between stages and pilot platforms - not as "grant scatterings" with weak traceability of effects.

Hence, the actors of science are interpreted not as a list of positions but as regimes of acceptance. Scientific "markets" are not a metaphor for "whom else to tell" but rather a multisided infrastructure of exchange - where the *currency* is evidence, its verifiability and compatibility, and the *price* is the cost of implementation and the time horizon to effect.

The language we propose allows us to design not only the content but also the fate of results: how they will be received, where they might get stuck, and how to help them reach practice.

We therefore maintain that this is precisely the task of *Scientomarket*: to transform fragmented communications into managed transitions between institutional worlds and to keep those transitions within the zone of ethical causality.

Chapter 3. Science as a Product

If the word *product* sounds somewhat jarrier when applied to science, it is only because it highlights aspects that academic culture often prefers to leave at the periphery - namely, specification, packaging, the user context, and the unit of exchange.

Scientific knowledge lives not only in the realm of ideas, papers, and arguments but also in a sociotechnical environment where it is recognized, tested, transferred, combined with existing practices, standardized, and ultimately implemented. Within this environment, a *scientific product* is understood not as a commodity in the narrow commercial sense but as a structured ensemble of claims and evidence, accompanied by reproducible artifacts (data, code, protocols, prototypes), rights of use, and interfaces that enable interaction with other bodies of knowledge and systems.

This perspective is necessary not for commercialization for its own sake but for the managed transition from value to trust, from trust to influence, and finally to sustainable resources that flow back into science to improve its quality. This distinction is crucial because, in some scientific and innovation systems, commercialization occurs for the mere sake of commercialization - that is, the global importance of translating knowledge into value is reduced to the bureaucratic plane of *reporting* and *accountability*.

Let us begin with the core.

We define a scientific product as a linkage between cognitive content (theoretical or empirical novelty) and operational form (which allows others to reproduce, verify, extend, and apply the result). Clearly, cognitive content without an operational form degenerates into rhetoric, whereas an operational form without cognitive content degenerates into empty technology.

To avoid both extremes, we interpret the product as a multilayered construct:

1. The inner layer is the *cognitive core*: formulation of the problem, model, hypotheses, evidence or data, and logic of inference,
2. The next layer is *verification*: experimental and analytical protocols, open datasets and code, replication procedures, and robustness and sensitivity reports,

3. The third layer is *translational*: methodological notes, adaptive instructions for related contexts (e.g., another industry, population, or hardware platform), and alignment with existing standards,
4. The fourth layer is *packaging*: versioning, metadata, identifiers, licenses, interfaces (APIs, data formats, documentation), compatibility control, and dependency management,
5. Finally, the rights-and-ethics layer includes the following layers: explicit licensing, disclosure of conflicts of interest, declarations of limitations, and boundaries of applicability.

All these layers must be coherent - otherwise, the product tears at the seams - during peer review, replication, or implementation.

In this framework, a minimal viable knowledge product (MVKP) is not merely a preprint. It is the minimal set of components that enables an external actor to verify and use the result. A typical MVKP includes the following:

- the corpus of claims (preprint or article),
- the corresponding dataset or at least a *synthetic dataset* of identical structure,
- executable code or well-documented pseudocode,
- procedures for reproducing key tables and figures,
- a file specifying the environment (container, dependencies), and
- logs (checksum files) verifying the integrity of the artifacts.

We may call this an evidence specification (ESpec) and a knowledge bill of materials (KBOM) by analogy with engineering, where the specifications and bills of materials define the reproducible integrity of products. The stricter the ESpec and the more complete the KBOM is, the shorter the path from *"read"* to *"reproduced"*, and therefore, the higher the knowledge conversion ratio between stages of the internal scientific cycle and the external markets of application.

Naturally, this raises again the question of product–market fit within the "knowledge market."

Conceptually, we have already expressed this through a simple function - formula (1)- which links evidential strength, utility for a given segment, time to effect, implementation costs, and risk.

The life cycle of a scientific product differs fundamentally from the classical "*research–publish–forget*" sequence. For us, it is cyclical and multistage-causal.

We distinguish seven stages:

1. Formation of the cognitive core: problem definition, model construction, and pilot data,
2. Internal verification: rigorous analysis, negative checks, sensitivity assessment, and protocol registration where appropriate,
3. Disclosure and verifiability: Preprint release, open data and code, a reproducible artifact package accompanied by ESpec/KBOM,
4. External validation: independent replications, extensions, and contextual transfers,
5. Translation and standardization: assembly of an *applied package* for industry or regulators, alignment with existing standards, and pilot implementations,
6. Implementation and feedback: contracts, licenses, practice changes, impact metrics, failure cases and their analysis,
7. Reinvestment: Part of the resources is directed toward strengthening reproducibility, expanding data, and refining theory - everything that, in the next cycle, increases E and reduces T.

This movement is conveniently monitored through the operational metric of the knowledge conversion ratio between stages, allowing us to see precisely where momentum is lost.

If N_k denotes the number of artifacts or cases that reach stage k and N_{k+1} denotes the number that successfully passes to the next stage, then

$$KCR = \frac{N_{k+1}}{N_k} \tag{2}$$

We believe that such a perspective obliges us to set engineering goals rather than rhetorical goals ("*we are actively popularizing science*"). For example, the knowledge conversion ratio (KCR) can be increased from 0.25 to 0.45 by containerizing the analysis, adding reference tests, and properly specifying metadata.

What we mean is that we are attempting to bring economics back into the conversation about the product. Every conversion has a cost; if the cost of transition exceeds the expected value for the target segment, the transition does not occur. We can express this as a logit-type probability of acceptance, where the parameters of utility and cost are governed by the architecture of the product and its interaction formats.

In general, without venturing too far afield, the unit of exchange in knowledge markets is not abstract "fame" but measurable indicators of trust and usability. Hence, we maintain that a scientific product must *carry* informal yet verifiable certificates of reproducibility, the results of independent replications, indicators of openness, disciplined version control, and resistance to contextual variation.

Culturally, this is not a simple shift because it demands reorientation - from *"announcing as soon as possible"* to *"verifying and applying as clearly as possible."* However, we believe that precisely in this change in assembly and maintenance practices lies the fastest gain in Science Product–Market Fit (SPMF). Sometimes, a single well-structured knowledge bill of materials (KBOM) can save months of correspondence and mistrust between a laboratory and its industrial partner.

Note that the portfolio of scientific products is rarely one-dimensional. From our own practice and other observations, we can see that laboratories benefit from constructing the following continuum of products:

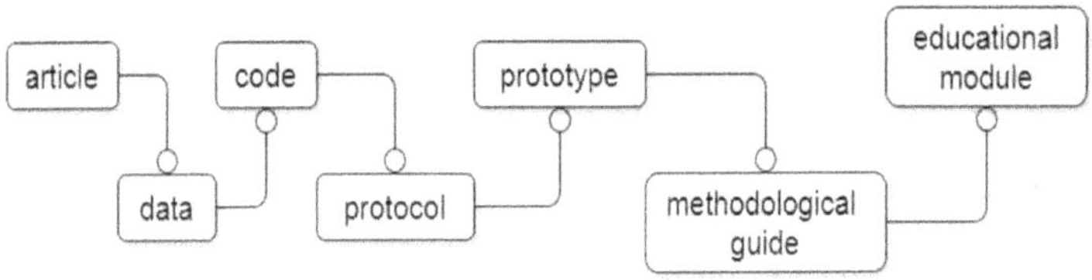

This coherent portfolio provides three main advantages.

First, it reduces friction among actors with different value criteria. Editors receive a rigorous paper; industry receives a demonstrative package; regulators receive a *policy brief* and a *standard-ready* application; and education receives a training set.

Second, it creates a multichannel reception effect: if the product does not yet reach one segment through a particular channel, it may still pass through another, thereby increasing the overall probability of adoption.

Third, it strengthens causality. An intervention in one layer (for example, adding reference tests to the code) increases trust in the entire package - and, consequently, in theory, the data and the protocol.

A scientific product does not exist in a vacuum. It is either interoperable or not. Its ability to enter existing technological and organizational chains depends not only on scientific correctness but also on its grounding in the "languages of interaction." In this way, we mean data formats, APIs, ontologies, identifiers, version policies, legal licenses, privacy and ethical regimes, and security requirements.

We may assert that any attempt to ignore these "low-level" issues proves (or can prove) costly at the translation stage. Hence, a mature scientific product must always contain—or at least aim to contain—a layer of alignment and interfacing with existing standards. Where no standard yet exists, the product should openly propose a "candidate standard" - that is, a minimal but sufficient specification around which compatibility can be built.

At this point, it is also appropriate to speak of the economics of the product.

The cost of creating and maintaining a product is distributed among its layers:

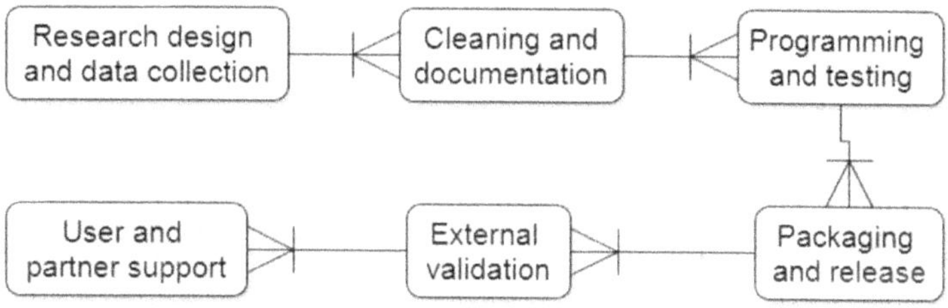

These expenditures are not redundant; they reduce future transaction costs and accelerate funnel transitions.

If we denote the total cost of creation and distribution as C and the integral value of impacts as V (including grants, agreements, implementations, and confirmed improvements in segmental metrics), then the knowledge return on investment (*K-ROI*) can be expressed simply as:

$$K\text{-}ROI = \frac{V - C}{C} \qquad (3)$$

Although formula (3) is trivial in form, it disciplines portfolio decisions. Sometimes, it is less expensive to bring a single product to the state where it "penetrates all segments" than to multiply fragile artifacts that fail every serious verification.

Conversely, bifurcating a product into two streams—one strictly academic and one industrial - with a clearly defined fork of interfaces and formats rather than trying to satisfy everyone with a single document may be more effective.

Finally, we must not forget the ethical dimension.

A scientific product must be transparent about its boundaries and uncertainties. This applies not only to papers but also to all other layers:

- to code, where there may be edge cases or hardware dependencies,
- to data, where sample biases may occur,
- to protocols, where hidden environmental parameters may be opaque,
- and prototypes, where safety concerns remain unresolved.

This ethical shell - our ethical marketing envelope (EME) - requires clear labeling of status (hypothesis, preliminary result, confirmed effect, implemented practice), avoidance of declared causality where only correlation is available (narrative causality risk), and transparency of conflicts of interest and funding sources. Indeed, a scientific product that “promises” more than it can demonstrate or support operationally destroys not only its own reputation but also undermines the institutions of trust upon which the knowledge market depends.

Thus, our interpretation of “science as a product” is neither the privatization of knowledge nor a marketing wrapper around emptiness. For us, it is above all the design of the full trajectory of a result:

from idea to verifiability,

from verifiability to transferability,

from transferability to compatibility,

from compatibility to implementation,

from implementation to feedback and reinvestment.

It is the readiness to think not only in terms of the truth of assertions but also in terms of the fate of assertions beyond their original laboratory.

Ultimately, the true product of science is not merely the article but an entire system of evidence and interfaces that allows other minds to act upon it.

Chapter 4. The Architecture of Value in Science

4.1. Scientific Value: A Working Definition

By *scientific value,* we do not mean the market price of knowledge or the social or scientometric prestige of its author. Rather, we refer to the capacity of a result to reduce uncertainty, expand the space of explanation, and become embedded in the reproducible practices of other researchers.

In our interpretation, scientific value manifests when a hypothesis becomes a *testable construct*, a protocol becomes a *transparent procedure*, data become an *accessible and documented resource*, and an interpretation becomes *robust to alternative explanations*[34].

Temporally, scientific value is long in time: its scale often unfolds only later, through independent replications, secondary analyses, and unexpected applications[35].

Moreover, of course, it is collective in nature - it becomes a public good only when the result moves beyond its original circle of authors and becomes anchored in infrastructures such as data repositories, code libraries, curricula, and methodological standards[36].

4.2. How Scientific Value Differs from Economic Value

Economic value seeks a single scale of measurement – price - that enables the comparison of goods and decisions under conditions of scarcity. In contrast, scientific value is inherently multidimensional and resists reduction to a single number.

Citations, impact factors, downloads, and other metrics are interpreted by us only as digital traces of circulation, not as indicators of the epistemic solidity of a finding.

---

[34] Merton, R. K. (1973). The Normative Structure of Science BT - The Sociology of Science. The Sociology of Science, (13), 267–278. Retrieved from https://press.uchicago.edu/ucp/books/book/chicago/S/bo28451565.html

[35] Leonelli, S. (2018). Data-Centric Biology. Data-Centric Biology. University of Chicago Press. https://doi.org/10.7208/chicago/9780226416502.001.0001

[36] Bornmann, L., & Daniel, H. D. (2008). What do citation counts measure? A review of studies on citing behavior. Journal of Documentation, 64(1), 45–80. https://doi.org/10.1108/00220410810844150

Economic valuation may rely on expectations and fashions, whereas scientific valuation requires procedural coverage: open materials, correct identification, replications, and critical dialog.

The carriers of economic value are primarily contracts and assets, whereas the carriers of scientific value are protocols, standards, public repositories, and educational practices.

In this sense, we argue that scientific value is closer to a public good than to a private asset because it increases through dissemination rather than depreciates by it[37].

4.3. The Conversion of Value: From Idea to Influence

For analytical clarity, we can distinguish a canonical chain through which most research results pass:

Each segment of this chain can be interpreted as follows:

Idea.

Value originates in the formulation of the question and the framing of the problem such that it becomes *solvable*. Here, clarity of definitions, explicit assumptions, and the ethical integrity of the planned procedures are crucial[38].

Publication.

This is not the finish line but a routing node. The material must be verifiable and transferable, with sufficient methodological description, access to data and code, and clear visualization[39].

---

[37] Ziman, J. (2002). A response to reflections on 'Real Science: What it is and what it means.' Science and Engineering Ethics, 8(2), 253–255. https://doi.org/10.1007/s11948-002-0026-1

[38] Popper, K. (2005). The logic of scientific discovery. The Logic of Scientific Discovery (pp. 1–516). Taylor and Francis. https://doi.org/10.4324/9780203994627

[39] Penrose, J. (2007). Beautiful evidence. Journal of Business Communication, 44(4), 427–434.

Citation.

Citations typically trace the *trajectories of circulation* across disciplines. They are indeed important, but they reflect interest and network effects rather than confirmed truth[40] (Garfield, 1955; Bornmann & Daniel, 2008; Fortunato et al., 2018).

Trust.

Institutionally secured confidence arises through peer review, openness of artifacts, independent reproductions, ethical policies, and sound error management.

Influence.

Influence manifests as a real reconfiguration of practice: when methods enter textbooks and standards, when data become benchmarks, and when code becomes part of widely used libraries - that is, when research agendas themselves are transformed.

In our framework, the key element of this scheme is feedback. Peer reviews, replications, methodological syntheses, and educational integrations return the result to a renewed examination of its methods and interpretations, thereby strengthening trust.

Feedback loops are, arguably, the most powerful mechanism for increasing both scientific value and epistemic trust.

Recently, we submitted a paper to a Q1-ranked Scopus journal (percentile > 90%). The authentic comments of four reviewers allowed us to significantly improve not only the scientific value of the work but also our own understanding of the problem. The second round of review led to an even greater increase in scientific value and, most importantly, to a deepening of mutual trust—when all participants in the process ultimately came to believe in the validity and meaning of the paper's results.

This is what we call a "*view from outside*" - the foundation of an objective perception of scientific value and trust, interpreted through the paradigm of positive feedback.

---

[40] Garfield, E. (1955). Citation indexes for science. Science, 122(3159), 108–111. https://doi.org/10.1126/science.122.3159.108

4.4. Differentiation of Scientific Value Types

To avoid reducing a rich and complex reality to a single metric, we distinguish four complementary types of value.

Epistemic Value

This represents the very core of scientific significance - the extent to which a result genuinely reduces uncertainty and increases the explanatory power of a theory. Indicators of epistemic value include sound identification, transparent statistics, accessible materials, successful independent replications, and robustness to alternative explanations.

Symbolic Value

By this, we mean the capacity of a result to become a discursive anchor - for instance, a named effect, a recognizable graph, or an eponymous method.

Symbolism, in this sense, facilitates communication and accelerates diffusion but becomes dangerous when it outpaces epistemics[41].

Institutional Value

Here, we refer to the degree to which a result changes the rules of the game - by shaping standards, launching consortia, influencing regulation, or becoming infrastructure (registries, benchmarks, libraries).

Institutional value anchors knowledge organizationally and makes it transferable.

Communicative Value

This is the capacity of a result to be understood, correctly applied, and reproduced outside the authors' circle - for instance, through clear language, robust visualization, intuitive navigation of artifacts, or educational materials[42].

Here, we mean that communication should not be seen as *cosmetic* but rather as the protective layer of epistemics.

---

[41] Whalley, P., & Gieryn, T. F. (2001). Cultural Boundaries of Science: Credibility on the Line. Contemporary Sociology, 30(1), 51. https://doi.org/10.2307/2654345
[42] Tufte, E. R. (2006). Beautiful evidence. Graphics Press, p.213, ISBN 0961392177

Evidently, the vector and balance of these four components largely determine the speed and completeness of conversion - from early attention to lasting influence.

4.5. The Scientomarket Value Chain

We find it necessary to introduce the term Scientomarket Value Chain to denote the organized sequence of nodes and channels through which an individual contribution is translated into public value, then into trust, and ultimately into influence.

This concept resonates with Porter's idea of a *"value chain"*[43], yet we apply it to science as a market of meanings, where the "links" are interpreted as productive, validating, transporting, institutional, and educational nodes.

Specifically:

Productive nodes: research groups, laboratories, data consortia, and computational centers.

Validating nodes: journals and platforms with transparent peer review, replication initiatives, and postpublication review communities,

Transport nodes: repositories of articles and preprints, protocol registries, code libraries, metadata and FAIR standards,

Institutional nodes: funding agencies, universities, professional associations, editorial policies,

Educational nodes include textbooks, courses, summer schools, and methodological guides.

The function of this chain is to reduce "*conversion throttles*" - that is, the points where strong results become stuck owing to closed data, weak methodological descriptions, lack of replications, or gaps between research and education (Hicks et al., 2015; Wilkinson et al., 2016).

[43] Porter, M. E. (1985). Competitive Advantage. Creating and sustaining superior performance. New York: The Free Press. Blood.

4.6. Where Value Is Lost: Typical Breakpoints

We can now outline the typical discontinuities that lead to the dissipation of scientific value:

1. *Methodological gap,*

   A strong idea but weak identification or incomplete procedural description. Interest exists, but trust is low,

2. *Data without FAIR compliance,*

   A solid piece of work whose data or code remain closed or undocumented. Such cases automatically hinder replication and slow the conversion into trust,

3. *Communication gap,*

   An epistemically strong result is presented so opaquely that it is difficult to understand or apply - leading to misinterpretation and reduced transferability,

4. *Institutional gap,*

   The findings are confirmed and clearly communicated, but a lack of standards or infrastructure keeps their influence local.

4.7. Normative Implications: Managing the Value Chain

Finally, we can outline practical recommendations for different actors on how to manage scientific value chains—which exactly should be done.

For authors:

The work is designed as a "value package" (text + method + data + code + replication guide + limitations). Plan for transferability to adjacent disciplines from the outset,

For editors and platforms:

Separate early popularity from verified validity. Label degrees of openness and replication status, support registered reports and postpone publication review formats,

For funders and universities:

Investments in replications and data infrastructures and recognition of contributions to standards, libraries, registries, and education are legitimate components of *impact*.

In summary, scientific value, as we interpret it, is neither an abstract virtue nor a sum of "*pluses*" in metrics.

It is the architectural capacity of knowledge to become a collective asset without losing its truth content.

Within this notation, the Scientomarket Value Chain translates that capacity into a manageable process - from a compelling idea to a reproducible publication, from early attention to confirmed trust, and from local success to structural influence.

Chapter 5. Trust as Scientific Currency

5.1. Trust as Currency

If in economics, exchange is mediated by currency, then in science, it is mediated by trust.

In this context, we understand trust not as a vague psychological state but as institutionally guaranteed confidence in the correctness of procedures, the integrity of authors, and the reproducibility of results.

Trust, in effect, translates the individual conviction of a researcher into the collective knowledge of a community, thereby ensuring the *liquidity* of scientific claims.

That is, the results circulate among laboratories, disciplines, and countries, becoming suitable for verification, extension, and application.

Importantly, we interpret trust not as "belief" but rather as a procedural infrastructure.

It is *built and maintained* through a set of concrete practices - from peer review and editorial standards to open data, preregistration, postponement review, and replication.

In this sense, trust functions as a universal medium of exchange: without it, value cannot be converted into influence, and influence degenerates into media noise.

5.2. Mechanisms of Trust

Peer Review: Risk Insurance at the Gate

In its strongest configuration, peer review is not a "taste filter" but rather a procedure for strengthening evidence.

It includes the assessment of research design, identification, statistical transparency, interpretation of limitations, and completeness of materials.

Modern modalities - double-blind review, open review, and registered reports - have shifted the emphasis from *"catching an effect"* to *"designing correctly."*

These formats mitigate publication bias and reframe review as a collective form of epistemic due diligence.

Editorial standards: The architecture of admission

Editorial boards and publishing platforms transform "good practices" into mandatory standards: access to data and code, methodological checklists, ethical approvals, preregistration statements, default open-materials policies, and dedicated space for replication notes.

When these standards are stable and transparent, the conversion of value into trust proceeds faster and with fewer distortions.

Indexing and Infrastructure: Maps of Trust Accessibility

Indexes and repositories are not merely catalogs - they are maps of accessibility.

They render results findable, comparable, and verifiable (Larivière, Haustein, & Mongeon, 2015).

The best emerging practice is the explicit labeling of trust status within each publication record: whether data and code are open, whether preregistration was completed, whether reviews are available, and whether independent replications have been published.

Such visible metadata create a grammar of credibility, enabling readers, editors, and funders to navigate not only results but also *their reliability gradient.*

Reputation: The Long-Term Credit Rating

By reputation, we refer to the historical record of engagement with uncertainty and error—how authors and journals correct inaccuracies, share materials, and respond to criticism.

In mature ecosystems, faultlessness is less valued than accountability. Reputation accrues through timely corrections, transparent amendments, and readiness for independent audits.

Thus, reputation operates as a long-term credit score of epistemic integrity - a cumulative history of how responsibly one handles being wrong.

5.3. The Erosion of Trust

Open access, in our view, is a historic gain - it removes barriers and accelerates exchange[44].

However, its rapid expansion has coincided with the general acceleration of the publishing industry, thereby exposing several systemic vulnerabilities:

1. *Speed over scrutiny,*

   Pressure on deadlines and volume has led to superficial reviews and "checkbox" procedures,

2. *Predatory journals,*

   The well-known phenomenon of *imitated legitimacy* - fictitious editorial boards, "turbo-reviews" - directly undermines the clearinghouse of trust, replacing genuine mechanisms with their formal facades.

3. *Publication distortions,*

   Publication bias, weak reporting of negative results, and "overfitting hypotheses" have collectively increased the fragility of the scientific corpus.

From this, we can draw a central conclusion: the problem lies not in open access itself but in unsynchronized institutions.

The remedy, therefore, is architectural.

We envision solutions in the certification of journals on the basis of real trust practices, cross-cutting checklists, funding of independent replications, visible trust statuses in article metadata, and education in the "grammar of trust".

[44] Tennant, J. P., Waldner, F., Jacques, D. C., Masuzzo, P., Collister, L. B., & Hartgerink, C. H. J. (2016). The academic, economic and societal impacts of Open Access: An evidence-based review. F1000Research, 5. https://doi.org/10.12688/f1000research.8460.1

Only by such systemic alignment can the scientific ecosystem maintain both velocity and veracity - the twin currencies of knowledge in motion.

5.4. Scientific Credit and Scientific Debt

Trust in the scientific ecosystem functions much like credit - a provisional loan of confidence that the community extends to a result.

Naturally, such credit has its own term, interest rate, and collateral:

Term,

After publication, a result often receives short-term credit of trust: it is read, cited, and discussed while the community awaits independent confirmation,

Interest rate,

The bolder the effect and the riskier the methodology is, the higher the "interest rate" and the more intensive the ensuing scrutiny,

Collateral,

Open data and code, preregistration, and complete methodological documentation serve as collateral that reduces the community's risk as the "creditor".

If subsequent verification confirms the result, the credit is converted into a long-term asset.

Trust becomes institutionalized - through replications, standards, and textbooks - and the result acquires structural influence.

If, however, further scrutiny exposes fragility, a scientific debt accumulates. Resources must then be spent on corrections, abandoned hypotheses, and revisions to teaching materials.

This debt cannot be "written off" without consequence because it is distributed among authors, editors, and institutions that have tolerated gaps in trust mechanisms.

Hence, the ethics of responsibility is both simple and strict: to request credit is to acknowledge the obligation of repayment.

In this sense, honest limitations, the publication of negative results, and transparent corrections are not rhetorical modesty but rather a form of financial literacy in the economy of scientific trust.

5.5. The structure of trust and the trajectory of perception

Let us imagine two studies within the same field.

Study A is striking and quickly published. The text is accessible, the visuals are compelling - but the data and code are unavailable, the design is described only in general terms, and the journal does not practice open review.

Study B, in contrast, shows a moderate effect but is fully preregistered: the data and code are open and documented, all deviations from the protocol are reported, and the journal publishes peer reviews and provides space for replication notes.

Suppose that during the first 12 months, *A* enjoys higher citation counts - particularly in "fast" journals and preprints.

Its effect becomes fashionable, appearing in reviews and presentations. However, after 24–36 months, replication cycles begin to differentiate the two.

Replications of *A* prove difficult; clarifications and critical comments accumulate.

Replications of *B*, however, succeed: independent reproductions, methodological extensions, and integrations into courses and standards emerge.

Ultimately, in the mature phase, trust and structural influence migrate toward *B*, and its citations become more "*load-bearing*" - accompanied by the reuse of data or code and the inclusion of its method in teaching materials and libraries.

From this, several straightforward conclusions can be drawn:

1. Early citation counts reflect interest, not trust,
2. The architecture of trust - openness, preregistration, open reviews, and replication support - alters the trajectory of perception,

It transforms a fashionable topic into a durable instrument that evolves into standards and pedagogy.

5.6. Policies for Trust Liquidity

What should institutions do to enhance the liquidity of trust within science?

1. *Label trust statuses on publication records,*

   Icons such as *"data/code open," "preregistered," "open reviews," and "independent replications (successful or failed)"* reduce information asymmetry and save community time,

2. *Fund replications and data infrastructures,*

   Replication should count as a career contribution, not a side pursuit of enthusiasts,

3. *Certify journals on the basis of trust practices,*

   Evaluation should depend not on impact factors but on actual procedures: review transparency, statistical validation, open artifacts, and responsible error management,

4. *Teach the "grammar of trust",*

   Courses in scientific writing and methodology should include training in preregistration, FAIR data, reproducible visualization, and open peer review,

5. *Separate popularity from validation.*

   Journals and platforms should analytically distinguish early attention from confirmed reliability, incorporating structural signals of influence such as textbooks, standards, and software libraries.

We thus interpret trust as the currency of scientific life. It arises only where there is collateral in verification and accountability. Peer review, editorial standards, indexing, reputation, and open science are not auxiliary services but the monetary and credit policy of the scientific system.

Erosion of trust begins when the emission of publications outpaces their verification coverage. To strengthen trust, the architecture of procedures must be synchronized with the speed of exchange.

Within the Scientomarket framework, trust is not decoration but the mechanism of circulation itself. Without it, scientific value cannot become public value, nor can influence become structural.

Thus, the task of institutions is not only to count metrics but also to govern the liquidity of trust - to make verifiability visible, replications incentivized, errors corrigible, and education oriented toward reproducibility.

Only then can the market of scientific meanings remain honest, and its currency – trust -retains its stability.

Chapter 6. Influence as a Mechanism of Knowledge Circulation

The word "influence" has lived too long in the shadow of counters. It has been reduced to numbers - how many times an article was cited, how many users downloaded a preprint, how many mentions a post received - digits conveniently placed into reports.

However, influence is not arithmetic.

It resembles a trajectory of meaning - the real movement of methods, ideas, data, and standards through the nodes of the scientific ecosystem: from one laboratory to another, from one discipline to another, and from classrooms to practice.

Metaphorically, metrics leave footprints on the sand, whereas influence reshapes the riverbed.

Thus, the first step is to restore the original sense of the term: *impact* is not how often we were noticed but what changed because we were noticed.

If ten mentions lead to a new measurement protocol, a revised course syllabus, a code library, or the incorporation of a method into a standard, we may call that *influence.*

Everything else is merely the temperature of discussion.

Such a redefinition frees researchers from the obsessive "view-count race" and returns them to a simple question: *Through which channels does my knowledge travel, and what happens to it along the way?*

Here, we begin to see that influence is not a static status once assigned to a text but rather a temporal dynamic.

First, there may be a flash of attention and discussion; next, the first wave of independent replications occurs; later, the result is integrated into the toolkit of a neighboring field; and only afterward does it find its place in textbooks or regulations.

These trajectories can be fast or slow, linear or winding - sometimes retreating after critical replications, sometimes reawakening unexpectedly years later. However, in every case, influence should be understood as the movement of ideas, not numbers.

If we imagine the scientific system as an organism, then its influence is its energy flow.

The value represents the mass of a result, trusts the currency of exchange, and influences the stream that sets the mechanism in motion.

Its vitality is revealed in how ideas flow from one research practice to another—how infrastructures (repositories, journals, and learning platforms) accelerate the transfer and how translators between disciplines build bridges across conceptual borders.

At the laboratory level, influence resembles a craft process: someone adopts your preparation protocol, another adapts your script, and a third builds an instrument following your schematic.

At the community level, influence manifests as new hubs where topics and approaches converge and where disciplinary boundaries are reinterpreted.

At the level of science as a whole, influence appears as a shift in norms: from closed to open data, from "fast sensations" to preregistration and reproducibility, and from isolated papers to value packages—bundles that include methods, code, data, and teaching materials.

Our discussion of influence therefore requires distinguishing its levels.

At the social level, one observes knowledge moving beyond academia and transforming into practice - clinical guidelines, engineering solutions, educational programs, urban designs, or national policies.

This is a long cycle: between publication and implementation lies a layer of intermediaries - professional associations, regulators, continuing-education institutes, journalists, and science communicators.

Here, clarity of language, ethical sensitivity, and the ability to handle uncertainty so that nonspecialists understand the boundaries and risks of effects become decisive.

At the institutional level, influence manifests within science itself, visible in how the rules of the game evolve: when journals adopt transparent peer review; when registered reports gain priority; when open data and code become the norm rather than the exception; when replications earn formal recognition; and when university reporting includes criteria of structural influence - for instance, contributions to standards, libraries, registries, and teaching modules.

This influence does not flare - it accumulates like pressure does. It is difficult to measure by a single metric yet easy to recognize by the growth of infrastructures that outlive specific projects and authors. At the metascientific level - science's reflection on itself - influences how we redefine evidence, learn to work with error, rebuild epistemic ethics, and rethink the role of metrics. This influence sets the gravitational field of the system: it changes expectations in laboratories and journals, determining what will count as "good work" in the future.

When the community begins to value not only *novel effects* but also clean replications, not only *high impact factors* but also transparent protocols, not only *strong p values* but also open data, may call this the triumph of metascientific influence.

How, then, do ideas actually spread? Above all, through form - through the qualities that make them portable. Clear writing, solid visualization, careful assumptions, transparent limitations, and accessible and well-documented artifacts all turn an idea into a transportable object. However, the form works only with a network structure.

Scientific networks are heterogeneous: they contain dense clusters where knowledge circulates rapidly and loose fields where novelty emerges through sparse bridges. Influence rarely travels in a straight line - it favors weak ties: cross-departmental seminars, summer schools, joint hackathons, and interdisciplinary internships.

There, unexpected combinations arise, giving birth to new interdisciplinary hubs.

The digital environment has added another layer - observable trajectories: preprints, code repositories, open datasets, teaching notebooks, blogs, and mailing lists. Digital traces reveal not only conversations around ideas but also their reuse: repository forks, new releases, pull requests from external researchers, and reanalyses of others' data.

This is the substance of influence—not endorsement but appropriation and continuation. Metaphorically, such movement has rails and obstacles. The rails are the standards and protocols that lower the cost of verification and transfer: templates for method descriptions, visualization checklists, metadata conventions, roadmaps for experimental replication, and teaching kits with step-by-step instructions. The obstacles are the points where energy dissipates: closed or poorly documented data, "black-box" code, careless figures, vague phrasing, and conclusions overloaded with assumptions. Managing influence begins with an honest diagnosis of these resistance nodes.

Sometimes a single, carefully prepared teaching notebook or short reproducibility guide suffices for an idea to "jump" into an adjacent field. Sometimes it requires a coalition or consortium to maintain infrastructure - a library, registry, or benchmark dataset.

However, to perceive influence as movement, we need a map.

In this context, the Scientomarket Map prototype is envisioned as a multilayer visualization where nodes represent roles rather than individuals, and edges represent types of transfer rather than citations:

On one layer lie productive nodes - laboratories, research groups, data consortia,

On another, validating nodes - journals, reviewers, postpublication communities, meta-research groups,

On the third, transport nodes - preprint servers, repositories, package libraries, and protocol registries,

Nearby are institutional layers - funders, universities, associations, and regulators,

Then, educational layers - textbooks, courses, schools, methodological centers,

Finally, public nodes - media, professional practitioner communities, and citizen-science platforms.

Between them run various flows: discussion and review, reuse of code and data, standardization in manuals and protocols, educational integration, transfer into regulation and everyday practice.

Overall, the map must be dynamic - showing over time how, in the early months, media flows and preprints dominate; later, the layer of replications and methodological extensions grows; and finally, standards and textbooks emerge.

Layers of status - transparency, verification, transferability, and maintenance - allow us to see instantly where movement slows, whether open artifacts exist, whether a platform for independent validation is available, and whether an educational pathway for adoption has been designed.

This map serves multiple functions simultaneously.

For authors, it becomes a mirror, revealing which "valves" are closed and what to add to reduce friction.

For editors, it acts as a control panel, displaying not only bibliographic metrics but also the life cycle of a result - its validation and infrastructural traces.

For funders, it provides an investment argument, highlighting where bridges are missing and which nodes yield the greatest public value when supported.

For society, it offers a transparent window into how knowledge becomes practice - and why this transformation takes time.

Above all, we must remember that the influence we seek should be understood not as applause or fame but as long-term transformation.

It requires *working in the long run*: a culture of evidential discipline, patience with replication, recognition of infrastructural contributions alongside "heroic" discoveries, and continual learning to navigate a complex ecosystem.

We must remember humility: every idea benefits when presented as an invitation to collaborate, not as a final word.

Where openness and accountability prevail, influence becomes structural - methods enter standards, data become benchmarks, code becomes libraries, and courses become living rewrites.

Under this logic, science gains immeasurably from the shift "from metrics to meaning."

Counters do not disappear; they merely move to the second line, serving as early signals, indicators of attention. On the front line, we place maps of transformation: we look not at the height of bars but at the depth of adoption, not at the number of mentions but at the number of practices embraced.

This perspective changes motivation, disciplines institutions, and teaches science to measure what truly advances knowledge—its capacity for honest and rapid circulation of ideas.

Chapter 7. Scientific Marketing and Evidence Narratives

To "speak science" means constructing a mode of exposition in which the reader not only learns the results but also, quite literally, walks the path of proof with the author: seeing the problem, understanding the logic of method selection, engaging the data, observing robustness checks, grasping limitations, and, in the end, sharing the conclusion within the stated uncertainty.

Persuasion here is born not of grand words but of a transparent form - of textual structure, clean visualization, disciplined rhetoric, informative titles and abstracts, and deliberate logistics of links and artifacts (data, code, protocols). In other words, we interpret the language of science as the operational interface of trust. It should minimize the reader's cognitive load without simplifying the substance[45].

When we call this "scientific marketing," we do not mean PR, hyperbole, or broad public popularization. We are speaking about the strategic management of attention within the scientific community: how to conserve a reviewer's time, how to furnish colleagues with verifiable footholds quickly, how to lead the reader from question to evidence without sacrificing rigor, and how to make the work findable, auditable, and portable to other laboratories. In this sense, *marketing* is, for us, a discipline of form. The task is not to "sell" results but to tune channels so that the work is easy to verify and adopt.

The hardest part is to be convincing without losing rigor. At the level of the sentence, persuasiveness begins with managed syntax and the flow of "given–new" information: a sentence should open with what the reader already knows from context and close with what is new. A paragraph must sustain a single topic–comment thread. This is not style for style's sake; it is necessity - precisely what reduces misinterpretation and attention "jumps".

At the level of sections, the classic IMRaD logic helps - provided we treat it not as bureaucratic form but as a plot: from problem and hypothesis to design and materials, from primary findings to checks and bounds, and from the particular to the general[46]. Structured abstracts and clear section headings improve discoverability and

---

[45] Gopen, G., & Swan, J. (1990). The Science of Scientific Writing: Writing with the Reader in Mind. American Scientist, (November-December), 1–33.

[46] Day, R. a, & Gastel, B. (2006). How to Write and Publish a Scientific Paper (How to Write and Publish a Scientific Paper. Medical Physics. Retrieved from http://www.amazon.com/dp/0313330409

the "first contact" with a paper. What matters is not length but informativeness: a title should signal the object, method, or main finding without the click-bait riddle[47].

Visualization is a distinct layer of evidence. A good figure does not decorate text; it shoulders part of the argument. It shows comparisons, the structure of uncertainty, causal assumptions, and the state of data "before" and "after" filtering. The principles are well known: avoid chartjunk; use perceptual encodings where the eye is most accurate (position on a common scale beats area and area beats volume); label units and samples explicitly; and show raw points when possible[48]. Visual annotations - "*what to look at*" - save reviewers time and reduce the risk of cherry picking, especially when the figure is accompanied by the script that generated it.

In general, graphical abstracts have become more widespread; for example, at MDPI, a graphical abstract is often on the default checklist.

Strict presentation is also a social grammar: where to place cautious modality, how to note alternative explanations, and what hedges to frame an effect with. In scientific writing, hedging is not a weakness but a form of responsibility: it marks boundaries of applicability and calibrates expectations[49].

Another piece of this grammar is the moves of the introduction and literature review. Editors expect you to name the area, identify the "gap," explain why your design addresses the right part of the problem, and explain how your protocol differs from what is known[50].

In the Methods section, reporting standards (CONSORT[51], PRISMA[52], and STROBE[53]) provide discipline. They do not constrain originality; they render details comparable and testable, without which reproducibility is an illusion.

---

[47] Jamali, H. R., & Nikzad, M. (2011). Article title type and its relation with the number of downloads and citations. Scientometrics, 88(2), 653–661. https://doi.org/10.1007/s11192-011-0412-z

[48] Cleveland, W. S., & McGill, R. (1984). Graphical perception: Theory, experimentation, and application to the development of graphical methods. Journal of the American Statistical Association, 79(387), 531–554. https://doi.org/10.1080/01621459.1984.10478080

[49] Piller, I., & Hyland, K. (1999). Hedging in Scientific Research Articles. Language, 75(3), 631. https://doi.org/10.2307/417106

[50] Pufahl, I., & Swales, J. M. (1993). Genre Analysis: English in Academic and Research Settings. Language, 69(1), 223. https://doi.org/10.2307/416471

[51] Schulz, K. F., Altman, D. C., & Moher, D. (2010). CONSORT 2010 Statement: Updated guidelines for reporting parallel group randomised trials. Italian Journal of Public Health, 7(3), 325–332. https://doi.org/10.4178/epih/e2014029

[52] Moher, D., Liberati, A., Tetzlaff, J., Altman, D. G., Antes, G., Atkins, D., … Tugwell, P. (2009, July 1). Preferred reporting items for systematic reviews and meta-analyses: The PRISMA statement. PLoS Medicine. Public Library of Science. https://doi.org/10.1371/journal.pmed.1000097

[53] von Elm, E., Altman, D. G., Egger, M., Pocock, S. J., Gøtzsche, P. C., & Vandenbroucke, J. P. (2014). The strengthening the reporting of observational studies in epidemiology (STROBE) statement: Guidelines for reporting observational studies. International Journal of Surgery, 12(12), 1495–1499. https://doi.org/10.1016/j.ijsu.2014.07.013

Why do we call all of this "*scientific marketing*"?

Attention is the scarcest currency in scientific exchange. Convincing forms reduce the transaction costs of reading and verification: reviewers more easily see strengths and weaknesses, colleagues more readily reproduce analyses, and students more quickly integrate methods into their toolkit. Intraacademic marketing involves the tuning of entry points: where and how to find artifacts, how repository navigation works, and how versions are annotated. Preprints, open data and code, and preregistration - these are not merely about speed and honesty; they are about the logistics of attention. When everything is assembled into a value package (text + data + code + reproduction guide), the conversion of interest into trust accelerates.

For example, if you accompany the paper with supplementary materials under DOIs (e.g., Zenodo), you significantly raise the chance that an editor will send it to review rather than desk reject - a direct outcome of increased trust.

Here, the concept of a narrative of evidence can be introduced. This is not a literary plot but a sequence of verifiable steps that connects the question to the claim and shows precisely which forms built the bridge.

This narrative rests on several pillars:

First, a clear question can be refuted or confirmed under the stated assumption. Without this, every subsequent action loses sharpness.

Second, design and protocol - not merely *what* was done, but *why* in that way; what alternatives were considered; what was preregistered; what changed en route; and for what reasons.

Third, materials - where the data reside, in what formats, how metadata are documented, and what the "roadmap" to the code looks like. Here, the FAIR principles apply: findable, accessible, interoperable, and reusable.

Fourth, the visual layer - figures that reveal uncertainty structure and effect bounds, not just a "beautiful story".

Fifth, robustness checks - alternative specifications, sensitivity to assumptions, negative results. These translate interest into trust.

Sixth, there is an honest discussion of limitations and the context of transfer - where the result does not apply, where other data are needed, and what systematic errors may lurk[54]. Limitations should always be addressed.

The form is not neutral. Title, abstract, and visual abstracts function as threshold experiences. Overly generic titles are lost; click-bait titles repel experts and erode trust. Conversely, structured abstracts and clear keywords aid searches and retain the relevant audience. In clinical and biomedical journals, visual abstracts do accelerate initial diffusion, but the same requirement applies: no simplifications that change meaning - all caveats must be visible[55].

In social media, lab blogs, and newsletters, the narrative must remain scientific: no categorical claims beyond the data; direct links to materials; and explicit answers to critical questions. A "fast" channel is not an amnesty from norms[56].

We must also clearly distinguish scientific marketing from popularization. Popularization translates ideas outward for broad audiences with different aims and forms. Scientific marketing operates inside the field: it configures protocols, visuals, and rhetoric so that colleagues can verify and transfer the work into their practices. The first task expands public understanding; the second reduces professional verification costs and accelerates the circulation of knowledge without quality loss[57].

Hence, the dangers in scientific marketing are anything that masks uncertainty: hyperbolic titles, aggressive claims of novelty, "quiet" omission of alternative specifications, or relativistic conclusions "outside the data." Studies show that exaggerations in press releases and headlines seep into the literature and public understanding, undermining trust in a discipline[58].

For example, a reviewer of our Q1 paper, whose title included the phrase "*When Readiness Meets Value*," strongly recommended a more scholarly title to avoid hyperbole.

---

[54] Gelman, A., & Loken, E. (2014). The statistical Crisis in science. American Scientist, 102(6), 460–465. https://doi.org/10.1511/2014.111.460

[55] Ibrahim, A. M., Lillemoe, K. D., Klingensmith, M. E., & Dimick, J. B. (2017). Visual Abstracts to Disseminate Research on Social Media. Annals of Surgery, 266(6), e46–e48. https://doi.org/10.1097/sla.0000000000002277

[56] Priem, J., Taraborelli, D., Groth, P., & Neylon, C. (2010). Altmetrics: a manifesto. October. Retrieved from http://altmetrics.org/manifesto/

[57] Gross, A. G., Harmon, J. E., & Reidy, M. S. (2009). Communicating science: The scientific article from the 17th century to the present. Parlor Press, p. 280, ISBN-10 : 1602351201

[58] Sumner, P., Vivian-Griffiths, S., Boivin, J., Williams, A., Venetis, C. A., Davies, A., … Chambers, C. D. (2014). The association between exaggeration in health related science news and academic press releases: Retrospective observational study. BMJ (Online), 349. https://doi.org/10.1136/bmj.g7015

Attention management begins long before publication - with the design of a representative form. If visualization checklists, repository structure, file-naming schemes, code versioning, README, and licenses are planned in advance, the output is not merely a "paper" but a reproducibility kit ready for immediate use.

This is how many methodological libraries are built: a standard repository structure, a teaching notebook, a minimal dataset for checks, and a list of known limitations. Such a kit is the best marketing tool because it lowers entry costs and makes the work genuinely in demand.

The narrative of evidence is especially vital where the system is under pressure to move fast: preprints and rapid communications. Speed does not excuse superficiality. A preprint is an invitation to dialog and must carry the same artifacts and caveats as the final version. Otherwise, "marketing" becomes noise and trusts a lottery.

This includes the postpublication layer: responses to comments, addenda with new checks, and open reviews - all of this continues the narrative, showing how a claim withstands the headwind of critique.

In the end, to "speak science" is to think in terms of transfer. A text must answer not only "*what did we do?*" but also "how will others adopt it?" Accordingly, it is sensible to include guides within the article: brief "*how-to*" for reproduction, a "*what can go wrong*" section, a minimal teaching case, and links to ready-made templates. This design does not "dumb down" the work; it makes it distributable - and thus genuinely influential within the field.

In sum, scientific marketing is not the shadow of science but its logistics: it ensures the movement of ideas, saves collective time, and increases the share of results that truly become new practices. This is what consolidates trust - when form does not conceal content but makes it visible, testable, and usable.

For us, the Narrative of Evidence names the discipline of form in which every step serves a single end: to make the path from question to conclusion observable and repeatable. On this basis, trust grows.

In this sense, scientific marketing is not glossy but clear - a form that does not adorn truth but helps it become public knowledge.

Chapter 8. The Future of Science as a Market of Meanings

We can see that for a century, the scientific system lived under the metaphor of industry. Laboratories were likened to factories, journals to assembly lines, and research careers to productivity tracks. This metaphor helped build the infrastructure of mass science (from grants to libraries). However, now it prevents us from seeing the essential point: knowledge is not "produced" or "delivered" like parts. In contrast, it circulates, settles, and transforms, becoming a collective capacity of society to act and understand. When attention shifts from article output to trajectories of impact, science reveals itself as an ecosystem of meanings: a network of nodes and routes in which value becomes trust and trust into durable changes in practices, standards, education, and institutions.

For us, the move "from industry to ecosystem" entails a shift in language and rules. In language, this is a move from reporting arithmetic to maps of transformation. In rules, it is a move from issuance speed to the liquidity of trust. This is not a rejection of metrics but their subordination to meaning. Under this subordination, counters become early signals and route indicators, not end goals.

From this perspective, editors and funders cease to be "suppliers of publications" and become operators of meaning markets that reduce friction at key nodes - helping readers understand and verify more quickly, encouraging data and code producers, supporting replications, and introducing intelligible trust statuses. In this context, universities cease to be "diploma factories" and become transfer nodes: they teach not only facts but also the grammar of reproducibility, the ethics of handling uncertainty, and the skills of infrastructural work - from maintaining repositories to standardizing protocols.

We can already observe this shift in practice. Preprints have ceased to be mere "draft signals" and have become entry points to dialog. Authors receive early expert feedback; reviewers gain access to data and code; readers can verify and extend results.

Open data and the FAIR principles have made possible what a decade ago seemed utopian: portability of studies - reproducing them, reusing them, combining them with new datasets, and building infrastructures upon them. The "*Manifesto for Reproducible Science*" and the Registered Reports format redirected focus from "sensations" to design quality, cutting off the temptation to retrofit hypotheses to data and rewarding honest treatment of uncertainty. Together, these form elements of a new ecosystem - not a factory where "numbers are delivered" but a market where verifiable truth circulates.

For this ecosystem to be viable, protocols for the future - light in form and deep in meaning - are needed. We highlight four of them - working titles that are also architectural requirements for exchange:

First protocol is StRL (*Study–Replicate–Learn*).

The idea is straightforward: each study should be conceived and packaged as a loop. "Study" means planning a verifiable design and making artifacts portable. "Replicate" is not an afterthought but a preplanned trajectory with data, code, and a guide. "Learn" means closing the loop by feeding lessons back into the protocol, visuals, training, and standards. StRL disciplines time, turning impact from a one-off spike into a cycle of improvement - visible and manageable.

Second protocol *is VALIDUS (provisionally: Verification, Accessibility, Limits, Interpretation, Data/Code, Use, Stewardship).*

By this, we do not mean another box-ticking checklist but a map of responsibility: where and how claims are corroborated; how accessible and documented the artifacts are; where the boundary of applicability is drawn; how assumptions are explained; how data and code are organized; how realistic use cases are anticipated; and how maintenance is ensured. VALIDUS precisely surfaces the items that are most often lost in the race for "novelty": a metadata glossary, a teaching notebook, "*what can go wrong*" scenarios, code versioning, and a policy for corrections.

Third protocol *is Scientomarket*, the framework to which our short monograph is dedicated.

Its task is to stitch together the chain "*value - trust - impact*" and to give institutions a set of levers: trust and openness statuses; support for replications as career contributions; analytics of structural impact (standards, textbooks, libraries) rather than popularity alone; bridge programs between disciplines; and journal certification on the basis of trust practices rather than one-dimensional indicators. We posit Scientomarket as both a voltmeter and a regulator—it shows where the system's idea-energy is lost to friction and how to reduce losses.

Fourth protocol is CSE (*Credible Science Experience*).

This name captures the thought that doing science is an experience that is reproducible from reader to reader. The CSE asks the author to assemble everything that makes the experience compelling: a minimal dataset; clean, commented code; annotated visualizations ("*what to look at*"); a short teaching case; and an honest map of limitations. Together, they form a trust kit by which a colleague does not merely believe but walks the path - and therefore transfers the result into practice.

Taken together, these protocols sketch the profile of a transparent, self-assessing, and self-developing science.

Transparent - because the verifiability status is visible from the paper's "passport": are data and code open, was there preregistration, are reviews open, are there independent replications.

Self-assessing - because checking does not end at publication; it continues in the postpublication layer: addenda, responses to critique, meta-analyses, and repository updates.

Self-development - Lessons from the current workflow back into protocols, courses, standards, and infrastructures. In such a system, error is not a defeat but rather a fuel for the next turn.

We argue that building this future ecosystem requires making its ethics explicit. We interpret the scientific economy as an economy of responsible constraints.

First, transparency must be reasonable: open by default yet not blind to privacy, security, and lawful access limits. A simple rule helps: if you cannot share raw material, share the route, so an independent researcher can reach the same point by the described road.

Second, rewards must follow infrastructural contributions: data, codes, standards, courses, and replications should carry career value no less than "new effects".

Third, responsibility for trust must be shared: authors for the honesty of form; editors for real review practices; funders for supporting checks and bridges; and universities for teaching the grammar of evidence and reproducibility.

Finally, we must distinguish popularity from corroboration: fast signals of interest constitute an invitation to verify, not an indulgence exempting verification.

This ethics implies a new "*politics of time*." Industrial logic measured success by publication speed. Ecosystem logic measures time-to-verified transfer: how quickly an external group reproduced the analysis; when a teaching module appeared; after how many months a method entered a protocol; and when data became reference grade. These intervals are not always shorter - but they are truer, reflecting the real costs of converting knowledge into practice and thereby increasing the durability of impact.

Seen through the eyes of science's users (actors - students, neighboring researchers, reviewers, journalists, regulators) - the key experience becomes navigation: the paper's card as a passport with trust and openness statuses, links to artifacts, a short reproduction route, a teaching example, and a changelog.

In our notation, journals become venues of experience: reviews do not vanish behind the curtain but become part of the scientific text; replications and negative results are not a shameful footnote but a normal part of the cycle; editorial decisions are backed by clear standards. Repositories become civic warehouses - not dumping grounds but places where data and code live, are updated, and are used; where there are maintainers and version responsibility; where citing software and data is not a courtesy but a norm.

Universities act as navigators: they teach how to read trust statuses, interpret visualizations, and distinguish early signals from established results. They build bridges across departments and practices - from seminars to hackathons and from schools to professional communities.

This future has enemies - hurry without coverage, metrics without context, black-box code, closed data without explanation, and fashion posing as novelty.

However, it also has strong allies: researchers who put the honor of form above the race; editors who change processes rather than slogans; funders who invest in infrastructures and replications; universities that teach how to speak science; and professional communities that make "quiet" contributions (methodological, educational, software) visible and valuable.

We can sum it up in one sentence:

The science of the future is a transparent, self-assessing, and self-developing ecosystem of impact, where the circulating capital is not money but verifiable truth.

It measures itself not only by citations but also by adopted practices; it evaluates not only the effect of publication but also the fate of results in repositories, standards, and textbooks; it preserves not only headline-grabbing discoveries but also the infrastructures that turn discoveries into public goods.

Thus, we believe that science must reclaim its market - not the market of money but the market of meaning. A market where value is defined by transparency and reproducibility, trust by procedures and responsibility and impact by changes in practices and standards. In this market, openness and verification statuses are visible, infrastructures and replications are rewarded, and errors become fuel for improvement. Only under such conditions does knowledge remain a public good and a force capable of changing the world.

Conclusion

In this brief, thesis-driven monograph, we started from a simple, yet not entirely transparent, idea: science is not reducible to the production of texts any more than industry is reducible to the output of parts. We interpret science as an architecture of meaning circulation in which epistemic value is converted into institutionally guaranteed trust and trust into durable changes in practice, standards, education, and, more broadly, public action.

This distinction is not a rhetorical flourish but a managerial necessity. As long as we gauge the viability of the scientific system by the pace of publication issuance and the number of citations, we see only the wave on the surface. To glimpse the current underneath, we must fix the conversion chain - from idea, through verifiable form, to collective use and to the infrastructures that make this conversion swift, honest, and reproducible.

Overall, the perspective we propose under the name Scientomarket is an attempt to supply a language for such optics.

We construed the "market of scientific meanings" as an ecosystem of nodes and flows where the circulating capital is not money but verifiable truth: journals and preprint platforms, data and code repositories, communities of reviewers and replicators, educational and professional institutions, regulators, and public arenas.

In this ecosystem, attention becomes the scarcest currency, and trust becomes the common clearinghouse. Any "signal of success" must have coverage - in the form of openness and verification.

Hence, our practical program distinguishes early signals of interest from confirmed impact; retunes editorial processes toward visible verifiability statuses; recognizes infrastructural contributions (data, code, protocols, instructional materials) as first-class research outputs; and finally returns replications to the center of the career economy of knowledge.

The historical sociology of science long ago revealed that truth in science was not only a logical category but also a social-organizational category because its strength rests on norms, trust-building practices, and material carriers.

In recent decades, the lessons of metascience have been added: local incentives for "fast novelty" and surrogate metrics can distort the entire regime of evidence production, creating a situation in which signals of popularity displace substance and publication issuance outpaces the development of verification infrastructures.

In this context, the "reproducibility crisis" is no longer a mere anomaly but rather a symptom of unsynchronized institutions: we accelerated dissemination channels without simultaneously accelerating verification procedures or making them visible in the "passport" of each work.

We therefore interpret our turn to Scientomarket not as a "new fashionable theory" but as an attempt to integrate three reform tracks:

1. Priority of reality over reporting. Registered reports and open-materials policies shift value from post hoc results to design correctness, completeness of description, and readiness for independent verification,
2. Infrastructural responsibility. FAIR principles, code versioning practices, and metadata documentation make knowledge transfer a material process, not a hopeful wager for the "conscientious reader",
3. Metric modesty. Insist on the contextual reading of indicators and reject one-dimensional measures in favor of maps of transformation - from standards and textbooks to libraries and registries.

Our central proposal can be stated as follows: if science is a market of meaning, then its monetary policy consists of norms and procedures of trust, not PR campaigns.

On this basis, the notions of "*scientific marketing*" and "*narrative of evidence*" have strict meanings.

The former is not about flashy titles but about reducing the transaction costs of verification: clear text structure, careful visualizations, explicit limitations, roadmaps to data and code, and teaching notebooks.

The latter is not about eloquence but about transparent argumentation: the reader must see the bridge from question to conclusion and all its supporting beams (assumptions, protocols, sensitivity to alternatives). This design is not cosmetic; it is the mechanism that converts value into trust and trust into impact. Where form conceals content, trust inflation is inevitable; where form serves content, what spreads is not noise but instruments.

No architecture endures without ethics. The knowledge economy of the future must be an economy of responsible constraints: "open by default" while "explaining exceptions," recognizing the rights and risks of participants, and caring for privacy and security without arbitrarily obscuring the grounds of inference.

Ethical literacy is not an appendix to methodology but its precondition: it determines what data may be collected and shared, what level of anonymization suffices, and how to describe uncertainty without masking vulnerabilities or inflating promises.

Ethics also concerns the fair distribution of credit: infrastructural and replication contributions must be visible and rewarded; otherwise, the market of meanings will mimic a market of spectacular storylines.

Who is the *Scientomarket* agenda for?

For authors, as a project frame: conceive the work as a value package including protocols, data, code, and a learning pathway.

For editors and platforms, as a regulatory frame: introduce badges for openness and verification, publish reviews, support replications and negative results, and separate early interest from confirmed impact.

As an educational frame, universities teach the grammar of trust and the logistics of transfer.

For funders and agencies, as an investment frame, finance the "bridges and roads" - replication centers, methodological libraries, data consortia, and long-term maintenance services.

For society, as a communication frame, the difference between early signals and confirmed knowledge is visible, thereby protecting the space of science from cycles of disappointment and skepticism.

The protocols we mention StRL (Study–Replicate–Learn), VALIDUS, CSE (to be developed in detail separately) and the Scientomarket framework itself form a minimal technology of the future.

StRL turns research from a one-off event into a learning cycle. VALIDUS renders visible the map of responsibility for each publication. CSE transforms reading into a reproducible experience, thereby accelerating transfer.

Together, they remove the chief "throttles" in the value chain: closed or poorly documented artifacts, opaque reviews, invisible replications, and overreliance on surrogate metrics.

These protocols are discipline-agnostic because they speak to the form of knowledge circulation, not its substantive content.

We recognize that this program requires time and discipline. However, it has powerful resources - the accumulated body of metascientific knowledge and existing practices already proven effective - Registered Reports in a range of journals, open-data and open-code policies in leading outlets, FAIR infrastructures, the emergence of robust libraries and reference datasets, and the revival of methodological reviews and teaching notebooks.

These lines are not finished, but their future geometry is, in essence, clear: where the liquidity of trust rises, the cost of error falls, and impact becomes structural - it enters standards and textbooks rather than vanishing with the next wave of announcements.

One final point remains, without which the argument dissolves into sensible but partial proposals.

Science is a common enterprise, and common enterprises rest on agreements about rules. Our return to the market of meanings is not the privatization of truth but, on the contrary, its deprivatization: we make conversion trajectories visible, institutionalize responsibility for trust, and redesign rewards so that infrastructure and verification are central.

If we can hold to these bearings, science will remain what it ought to be: the collective capacity of the world to learn, to correct itself, and to act in the presence of uncertainty. Then, the formula we put on the cover will cease to sound like a slogan and become a description of reality: Science must reclaim its market - not a market of money but a market of meaning.

A market where value is defined by transparency and reproducibility, trust by procedures and responsibility, and impact by changes in practices and standards.

www.ingramcontent.com/pod-product-compliance
Lightning Source LLC
LaVergne TN
LVHW022314200926
848481LV00004B/1342
9798899662737